Independent Contractor Agreements

REVISED
SECOND
EDITION

*Ten Sample Independent
Contractor Agreements for a
Variety of Business Situations*

By James R. Urquhart III

Revised Second Edition.

First printing, February, 1989

Printed in the United States of America

International Standard Book Number:

0-094496-04-3

The Fidelity Publishing Corporation of America
2061 Business Center Drive - Suite 112
Irvine, California 92715-1107
Telephone: (714) 752-5544

TABLE OF CONTENTS

This page is intentionally left blank.

FORWARD

These form independent contractor agreements are representative of those that I have seen over the years from various sources. Since they cover a number of different business situations, they may be of some assistance to you or to your attorney in drafting your own agreement. A single boilerplate form to fit all situations does not exist and could not be created.

These agreements are not meant to replace your attorney. They have neither been tailored to your specific needs, nor do they necessarily reflect any special legal requirements that may exist in your state. "Cut and paste" provisions from these agreements as you see fit, but I urge you to spend a few minutes with your attorney before you put your own independent contractor agreement into actual use. As any lawyer knows who regularly prepares contracts for his clients, the so-called boilerplate agreement is only a starting point. Changes are almost always needed to conform the agreement to the particular facts facing that client and to the particular law of the jurisdiction at hand.

January, 1989

James R. Urquhart III
Irvine, California

This page is intentionally left blank.

DEDICATION

One morning I rose before sunrise. As usual, that day I knew that my schedule would be busy. The only way I was going to get any writing done was to get to the office before the rest of the staff arrived and the phones started to ring. I heard a gentle knock on our bedroom door. Before I could take a step in that direction, the door was quietly pushed open by my six year old daughter, Barbara. She had not yet learned to tell time. Whether or not it was light or dark outside was still her best way to tell time. Not wanting to wake her mother or her three year old brother Jimmy down the hall, she whispered,

> "Daddy, why are you getting up in the middle of
> the night?"

She must have heard my electric shaver or some other sound I had made. I was trying to be quiet, but, apparently, not quiet enough. I explained,

> "It's not the middle of the night. It's almost morning.
> I'm just going into the office early to work. Go back to
> bed. You don't have to be up for another two hours."

Still a little puzzled, she went back to her room and to bed.

This humble publication is, therefore, dedicated to my family, who have endured my absence on many occasions. More times than I like to think about, I have not been there when my children get up in the morning, nor when they go to bed at night. My wife Deborah keeps things together when I can't be there. I am thankful to her for that.

A PRIMER ON THE LAW OF
INDEPENDENT CONTRACTOR AGREEMENTS

Even a well drafted independent contractor agreement will not convert an employee into an independent contractor.

The independent contractor agreement represents the belief of the parties as to the nature of their relationship. While the belief of the parties to the contract is relevant evidence on the independent contractor or employee issue, belief or state of mind standing alone does not decide the issue. For example, A and B can agree in writing that A shall use B to perform certain services as an independent contractor and not as an employee. The courts, and the various government agencies that are interested in the issue, are not bound by such an agreement. The parties may be mistaken in their belief. Or, other relevant facts may contradict their self-serving declaration of independent contractor status.

The federal government with respect to employment taxes, as a general rule, uses the "usual common law" test to determine employee or independent contractor status. Currently, the Internal Revenue Service is using 20 factors that it feels accurately reflect the common law distinction between employee and independent contractor. The Service's most current restatement of those 20 factors may be found in Revenue Ruling 87-41 (1987-23 I.R.B. 7). The factors have been developed based on an examination of cases and rulings considering whether an individual is an employee. I have set forth the 20 factors in a footnote.[1] Note that belief of the parties or the existence of an independent contractor agreement is not even mentioned in the 20 factors. I feel that the Service is tired of hearing from a taxpayer - employer that the agreement _says_ that the worker is an independent contractor, therefore, the worker _is_ an independent contractor. To stop such an argument before it starts, belief of the parties and written contracts are left out of the 20 factors. There are in fact more than 20 factors. The Service has chosen to list the 20 it feels are more instructive on the issue.

The _Restatement of Agency_, _Second_, section 220, is an often referred to source for the common law on the independent contractor vs. employee issue. The _Restatement_ uses the ancient words "master" for employer, and "servant" for employee.

Many states, through their decisional law have adopted the _Restatement_ view of master and servant in connection with various issues such as liability of the employer for acts of the employee (a form of vicarious liability) and workers' compensation laws.

---

1 (1) Instructions; (2) Training; (3) Integration; (4) Services rendered personally; (5) Hiring, supervising, and paying assistants; (6) Continuing relationship; (7) Set hours of work; (8) Full time required; (9) Doing work on employer's premises; (10) Order or sequence set; (11) Oral or written reports; (12) Payment by hour, week month; (13) Payment of business and/or traveling expenses; (14) Furnishing of tools and materials; (15) Significant investment; (16) Realization of profit or loss; (17) Working for more than one firm at a time; (18) Making service available to general public; (19) Right to discharge; (20) Right to terminate.

The <u>Restatement of Agency, Second</u>, section 220, provides in part:

> "(1) A servant is a person employed to perform services in the affairs of another and who with respect to the physical conduct in the performance of the services is subject to the other's control or right to control.
>
> (2) In determining whether one acting for another is a servant or an independent contractor, the following matters of fact, among others, are considered:
>
> . . .
>
> (i) whether or not the parties believe they are creating the relation of master and servant; . . ."

What the parties "believe" is merely a state of mind. As stated above in section 220(2)(i), what the parties "believe" is just one of many "matters of fact" to be considered.

Comment "h" of section 220 states in part:

> "Factors indicating the relation of master and servant.
>
> The relation of master and servant is indicated by the following factors: an agreement for close supervision or de facto close supervision of the servant's work; . . . an agreement that the work cannot be delegated."

As can be seen from Comment "h", an independent contractor agreement that provides for "close supervision" or that the work "cannot be delegated" tends to show employee status, and not independent contractor status.

Work that can be delegated is work that can be assigned to someone else. In most cases, an employee is expected to do the assigned work personally. In contrast, an independent contractor can, and often does, delegate or assign the work to an employee of the independent contractor or to another independent contractor. The contract between the parties to an independent contractor relationship should expressly state that the work is assignable or delegable.

The <u>Restatement of Agency, Second</u>, section 220, Comment "m" states:

> "Belief as to existence of relation.
>
> It is not determinative that the parties believe or disbelieve that the relation of master and servant exists, except insofar as such belief indicates an assumption of control by one and submission to control by the other. However, community custom in thinking that a kind of service, such as household service, is rendered by servants, is of importance.
>
> Illustrations:
>
> 10. A, employed by a taxi company, is sent by P, his employer, to drive B from X to Y, and it is agreed between A, P, and B that for the purposes of the trip A is to be B's servant, although B is to exercise no more control over A's conduct than is normal in the ordinary case of passengers in taxicabs. A is not B's servant.
>
> 11. A is employed by P as resident cook for his household under an agreement in which P promises that he will in no way interfere with A's conduct in preparing the food. A is P's servant."

In conclusion, <u>in close cases</u> a well drafted independent contractor agreement <u>can</u> make the difference between denial or acceptance of the independent contractor relationship. In very clear cases of employee or independent contractor, whether or not there is a written agreement is of little importance. In very clear cases, the facts cause the scale to ground-out one way or the other. In such cases, a written contract between the parties would not materially change the balance.

This page is intentionally left blank.

A WORD TO THE WISE:
THESE CONTRACTS ARE NOT MAGIC

In the fairytale *Cinderella*, the good fairy turned a pumpkin into a shiny new carriage. The good fairy can work such magic. These independent contractor agreements cannot.

An employee will not be turned into an independent contractor through the use of so-called independent contractor agreements. Do not think, merely because a specific type of independent contractor agreement exists in this form book, that that type of worker is always an independent contractor. Whether a worker is an independent contractor or an employee, in the absence of a specific statute, is generally determined by an examination of all the facts and circumstances. For example, though one of the independent contractor agreements in this form book relates to couriers and messengers, there is no hard and fast rule that all couriers and messengers are independent contractors. They may or may not be depending on the given factual situation. Belief of the parties as reflected in their independent contractor agreement is just a single matter of fact to be weighed on the scales along with many other matters of fact.

ABOUT THE AUTHOR:
JAMES R. URQUHART III

Mr. Urquhart, a practicing attorney with law offices in Irvine, California, is an expert on the independent contractor vs. employee issue. He is the editor of <u>The Independent Contractor Report</u>, a monthly newsletter covering matters of importance to independent contractors and the businesses that use them. Over the years, throughout the United States, thousands have attended his seminar entitled <u>Independent Contractor v. Employee</u>. As a specialist actually practicing in the field, and as a nationwide lecturer on the subject, Mr. Urquhart brings a unique background to this complex and hazardous area of the employment tax law.

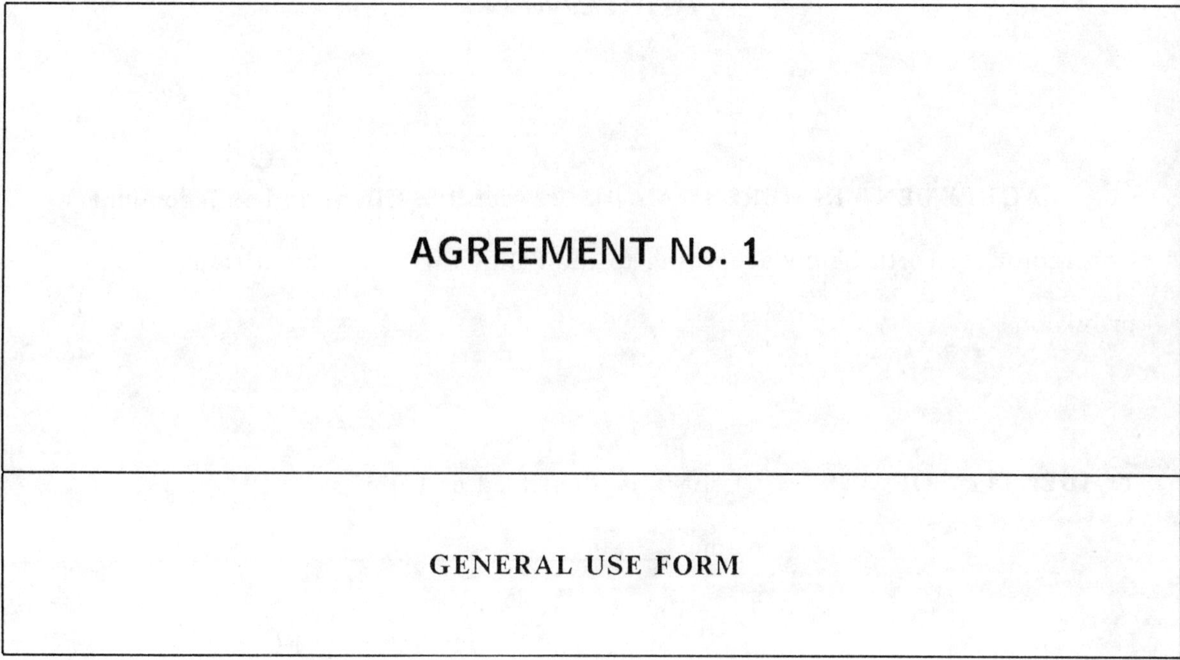

Comment:

This agreement was developed for general use. In other words, it was written with no specific situation in mind (for example, construction, carpet cleaning, or direct selling). Because of its very general nature you may find it easily adaptable for your business needs.

AGREEMENT

AGREEMENT IS HEREBY MADE between the Client and Independent
Contractor set forth below according to the following terms, conditions, and
provisions:

1. IDENTITY OF CLIENT

CLIENT is identified as follows:

Name: _______________________________________

Type Entity: [ ] Sole Proprietorship
                [ ] Partnership [ ] Corporation

Address: _____________________________________

City/State/Zip: ______________________________

Business Telephone: __________________________

2. IDENTITY OF INDEPENDENT CONTRACTOR - "IC"

The Independent Contractor (hereafter "IC") is
identified as follows:

Name: _______________________________________

Type Entity: [ ] Sole Proprietorship
                [ ] Partnership [ ] Corporation

Address: _____________________________________

City/State/Zip: ______________________________

Business Telephone: __________________________

Social Security or Employer Identification Number:
____

License Number and Expiration Date, if any:
____
____

| **3. WORK TO BE PERFORMED** | CLIENT desires that IC perform, and IC agrees to perform, the following work: |

____

____

____

| **4. TERMS OF PAYMENT** | CLIENT shall pay IC according to the following terms and conditions: |

____

____

____

| **5. REIMBURSE-MENT OF EXPENSES** | CLIENT shall not be liable to IC for any expenses paid or incurred by IC unless otherwise agreed in writing. |

| **6. EQUIPMENT, TOOLS, MATERIALS, OR SUPPLIES** | IC shall supply, at IC's sole expense, all equipment, tools, materials, and/or supplies to accomplish the work agreed to be performed. |

| **7. FEDERAL, STATE, AND LOCAL PAYROLL TAXES** | Neither federal, nor state, nor local income tax nor payroll tax of any kind shall be withheld or paid by CLIENT on behalf of IC or the employees of IC. IC shall not be treated as an employee with respect to the services performed hereunder for federal or state tax purposes. |

| **8. FRINGE BENEFITS** | Because IC is engaged in IC's own independent business, IC is not eligible for, and shall not participate in, any employer pension, health, or other fringe benefit plan, of the CLIENT. |

| **9. NOTICE TO IC REGARDING ITS TAX DUTIES AND LIABILITIES** | IC understands that IC is responsible to pay, according to law, IC's income taxes. If IC is not a corporation, IC further understands that IC may be liable for self-employment (social security) tax, to be paid by IC according to law. |

| **10. CLIENT NOT RESPONSIBLE FOR WORKERS' COMPENSATION** | No workers' compensation insurance shall be obtained by CLIENT concerning IC or the employees of IC. IC shall comply with the workers' compensation law concerning IC and the employees of IC. |

11. TERM OF AGREEMENT

This agreement shall terminate at 12:01 a.m. on ________________, 19______.

12. TERMINATION WITHOUT CAUSE

Without cause, either party may terminate this agreement after giving 30 days written notice to the other of intent to terminate without cause. The parties shall deal with each other in good faith during the 30 day period after any notice of intent to terminate without cause has been given.

13. TERMINATION WITH CAUSE

With reasonable cause, either party may terminate this agreement effective immediately upon the giving of written notice of termination for cause. Reasonable cause shall include:

    A. Material violation of this agreement.
    B. Any act exposing the other party to liability to others for personal injury or property damage.

14. NON-WAIVER

The failure of either party to exercise any of its rights under this agreement for a breach thereof shall not be deemed to be a waiver of such rights or a waiver of any subsequent breach.

15. NO AUTHORITY TO BIND CLIENT

IC has no authority to enter into contracts or agreements on behalf of CLIENT. This agreement does not create a partnership between the parties.

16. DECLARATION BY INDEPENDENT CONTRACTOR

IC declares that IC has complied with all federal, state and local laws regarding business permits and licenses that may be required to carry out the work to be performed under this agreement.

17. HOW NOTICES SHALL BE GIVEN

Any notice given in connection with this agreement shall be given in writing and shall be delivered either by hand to the party or by certified mail, return receipt requested, to the party at the party's address stated herein. Any party may change its address stated herein by giving notice of the change in accordance with this paragraph.

**18. ASSIGN-
ABILITY**

This agreement may be assigned, in whole or in part, by IC. IC shall provide written notice to CLIENT promptly before any such assignment.

**19. CHOICE OF
LAW**

Any dispute under this agreement or related to this agreement shall be decided in accordance with the laws of the State of _______________________________.

**20. ENTIRE
AGREEMENT**

This is the entire agreement of the parties and cannot be changed or modified orally.

**21. SEVER-
ABILITY**

If any part of this agreement shall be held unenforceable, the rest of this agreement will nevertheless remain in full force and effect.

22. AMENDMENTS

This agreement may be supplemented, amended or revised only in writing by agreement of the parties.

"CLIENT":

____
Name of Client

By: X_______________________________   Date:_______________

Its: _______________________________

"IC":

____
Name of IC

By: X_______________________________   Date:_______________

Its: _______________________________

AGREEMENT No. 2

TILE INSTALLATION SERVICES

Comment:

This independent contractor agreement deals with the construction business, specifically tile work. Personally, I don't care for the "First Party" and "Second Party" approach, but otherwise it is a well thought out agreement. Paragraph number 4 provides for instrumentalities to be supplied by the IC and paragraph 11 talks about the IC "holding himself out to the general public". These are good provisions to include in almost any IC agreement. I question the enforceability of the "hold harmless" clause in paragraph 7. It may be against public policy to allow an employer to sue a worker, held on audit to be an employee, for payroll taxes.

AGREEMENT

AGREEMENT BETWEEN ______________________________________

as FIRST PARTY and ______________________________________ **as**

SECOND PARTY:

1. WORK TO BE PERFORMED	SECOND PARTY shall perform all services described by FIRST PARTY'S work order in accordance with the drawings and specifications contained or incorporated therein.
2. TERMS OF PAYMENT	FIRST PARTY agrees, except as hereinafter provided, to pay SECOND PARTY for services performed by SECOND PARTY and accepted by FIRST PARTY in accordance with the terms of paragraph six of this agreement. Payments will be made weekly on all acceptable work. Work shall be deemed not acceptable if FIRST PARTY'S customer refuses to accept the work or the work is not done in accordance with prevailing tile and contractor's standards. FIRST PARTY shall have the right to withhold in an interest-bearing retention fund, ten percent of all sums due to SECOND PARTY, such withholding not to succeed $500.00. All such sums retained shall be returned after the expiration of eighteen months from the termination of this contract, reduced by the amounts of chargebacks or debts due from SECOND PARTY to FIRST PARTY, as specified in Paragraph 9 herein.
3. TERM OF AGREEMENT	This agreement is to become effective as of this date and is to continue in full force and effect for one year. It is mutually agreed that this contract shall automatically renew itself on each expiration year unless written notice is given by either party to the other at least thirty (30) days prior to expiration.

4. LABOR, SERVICES, SUPPLIES, EQUIPMENT, TOOLS

Except as otherwise provided herein, SECOND PARTY shall provide all labor, services, supplies, equipment and tools of every kind required for the prompt and efficient execution of the work described herein.

5. BUILDING MATERIALS

FIRST PARTY shall furnish to SECOND PARTY all required building materials. SECOND PARTY shall be liable for any loss or damage occurring to such materials while they are in the possession of SECOND PARTY. All materials not used on the job shall be returned to FIRST PARTY in reusable condition.

6. COMPENSATION, NO AUTHORITY TO BIND FIRST PARTY

FIRST PARTY shall compensate SECOND PARTY according to the following terms and conditions:

____

_______________________________________.

Said compensation shall be SECOND PARTY'S entire compensation, including all expenses incurred by him in the performance of this Agreement, and he shall have no power to incur any debts or other obligations on behalf of FIRST PARTY.

7. NOTICE TO SECOND PARTY REGARDING ITS TAX DUTIES AND LIABILITIES

FIRST PARTY NOT RESPONSIBLE FOR WORKERS' COMPENSATION

SECOND PARTY agrees to accept full responsibility for any and all taxes that may be lawfully due to any governmental unit and to hold the FIRST PARTY harmless from any liability for the non-payment of taxes due from SECOND PARTY to any governmental unit. SECOND PARTY waives any and all claim from FIRST PARTY to any form of workers' compensation insurance coverage or compensation provided under Federal, State or Local law which affects employees and employers, and agrees to carry and provide his own insurance for injury or sickness or retirement, whether in the form of Social Security or otherwise.

8. ASSIGNABILITY

Neither this agreement nor the right to any payments hereunder shall be assignable by SECOND PARTY and any purported sale or assignment at any time shall make the agreement null and void, and unconditionally forfeit all interest in the SECOND PARTY.

9. ACCEPTANCE OF WORK, CHARGEBACKS

Compensation shall be paid only on acceptable work, and FIRST PARTY may offset against any claims for compensation hereunder chargebacks for faulty work and any debts or monies due for damages by SECOND PARTY to FIRST PARTY. SECOND PARTY shall have the right to correct such faulty work at his own expense immediately upon notice from FIRST PARTY. If FIRST PARTY cannot, through the exercise of reasonable diligence, locate SECOND PARTY, or if FIRST PARTY'S customer refuses to allow SECOND PARTY to correct such work personally, FIRST PARTY may complete or correct such work in the manner in which FIRST PARTY sees fit and backcharge SECOND PARTY therefore.

10. MONIES RECEIVED, MANNER OF PERFORMANCE OF SERVICES

SECOND PARTY shall: (a) consider as trust funds belonging to FIRST PARTY any monies received as due or to become due FIRST PARTY and shall immediately deliver such monies to FIRST PARTY; (b) not perform services in any manner which will bring discredit or injury to FIRST PARTY or employ any misrepresentation or scheme in the performance of services.

11. HOLDING OUT TO GENERAL PUBLIC

SECOND PARTY states and affirms that he is acting as a free agent and independent contractor, holding himself out to the general public and maintains his office and principal place of business at own address, and that this Contract is not exclusive. Except as otherwise specified herein, FIRST PARTY possesses no right hereunder to discourage or inhibit the SECOND PARTY'S rights to enter any other contracts as SECOND PARTY sees fit.

12. SOLICITATION OF FIRST PARTY'S CUSTOMERS

SECOND PARTY hereby agrees that within a period of one year from the date of last performing work under this Contract, he will not solicit or perform or aid any other person in the solicitation or performance of any work of a similar nature to the work performed by FIRST PARTY for any of FIRST PARTY'S customers in the City of ______________________, State of ______________________________ and twenty-five mile surrounding area.

**13. LIABILITY
FOR CLAIMS AND
LIENS**

SECOND PARTY shall at all times indemnify and hold
FIRST PARTY harmless against all liability for claims
and liens for labor performed and materials used or
furnished to be used on the job. SECOND PARTY
agrees within ten (10) days after written demand to
cause the effect of any suit or lien to be removed from
the premises and, in the event that SECOND PARTY
shall fail to do so, FIRST PARTY is authorized to use
whatever means in its discretion it may deem
appropriate to cause said lien or suit to be removed or
dismissed, and the costs thereof together with
reasonable attorneys' fees shall be immediately due and
payable to FIRST PARTY by SECOND PARTY.

**14. CLAIM,
LIABILITY, LOSS
OR DAMAGE AS A
RESULT OF
SECOND PARTY'S
NEGLIGENCE**

SECOND PARTY shall indemnify and hold FIRST
PARTY harmless from and against any claim, liability,
loss or damage, including reasonable attorneys' fees,
arising by reason of the death or bodily injury of
persons, injury to property or other loss or damage
resulting from SECOND PARTY'S alleged or actual
negligent act or omission.

**15. REMOVAL OF
WASTE MATE-
RIALS**

During the course of construction, SECOND PARTY
shall remove waste materials from the site as necessary
to maintain the premises in a clean and orderly
condition. Upon the completion of the work, SECOND
PARTY shall remove all debris and waste incident to
the operation and clean all service fixtures and
equipment relative to the performance of this
Agreement.

**16. GUARANTEE
OF WORKMANSHIP**

SECOND PARTY guarantees all workmanship and
agrees to replace at his sole cost and expense, any and
all workmanship adjudged defective or improperly
installed for a period of three years from completion
and acceptance of the work performed under this
Contract.

**17. PREVIOUS
CONTRACTS,
AMENDMENTS**

This shall serve as the only Agreement between FIRST
PARTY and SECOND PARTY. All previous written or
oral contracts are hereby canceled. Any amendments
hereto must be in writing and signed by both SECOND
PARTY and an officer of FIRST PARTY.

18. TIME   Time is of the essence of this contract.

IN WITNESS WHEREOF we have affixed our signatures this

_______________ day of _______________________, 19_________.

____
By:
FIRST PARTY

____
By:
SECOND PARTY

<u>CAUTION</u>: This form is an illustration only. It has not been prepared with any specific factual situation in mind. It may not be suitable for your particular use. No warranty is made as to its fitness for any particular purpose or use. No other warranty is given. Before using this agreement, or any part of it, you should consult with your attorney.

AGREEMENT No. 3

ESTIMATING SERVICES

Comment:

This independent contractor agreement relates to estimating, bidding, sales, and marketing services. I like paragraph 5 whereby it is stated that the agreement shall not be construed as a partnership. From time to time contentions of partnership do arise unless specifically eliminated. If someone contends they are your partner, they might want to share in your profits and the partnership property. Paragraph 3.2 on reimbursement is not the best idea. Independent contractors should pay their own expenses without reimbursement.

AGREEMENT

This Estimating Services Agreement ("Agreement"), effective ________________,
19_____, is made by and between ________________ (hereinafter referred to as
"Client"), and ____________________ (hereinafter referred to as "Contractor"). In
consideration of their mutual promises and for their mutual benefit, Client and
Contractor agree as follows:

1. APPOINTMENT AND DUTIES OF CONTRACTOR

Client hereby appoints Contractor to perform estimating, bidding, sales and marketing services and related services for ____________________ under the terms and conditions set forth herein. Contractor accepts such appointment and agrees to work diligently and to use its best efforts in connection therewith subject to termination as provided in Section 4 below.

2. CONTRACTOR

Contractor represents that it is qualified to perform the duties set forth in Section 1 above.

3. FEES

3.1 Monthly Fee. Client shall pay to Contractor a fee of $________________ for each complete month while this Agreement is in effect. Said payment will be prorated for any period this Agreement is in effect which is less than a complete calendar month.

3.2 Reimbursement for Certain Business Expenses. Client will also reimburse Contractor for all reasonable meal and entertainment expenses incurred by Contractor in performing its services pursuant to this Agreement. Contractor

must provide adequate evidence of any meal and
entertainment expenses prior to reimbursement by
Client. Client retains the right to place a limitation on
the amount of meal and entertainment expenses
Contractor will be reimbursed for at any time while
this Agreement is in effect, provided that Client
notifies Contractor in writing and such limitation
applies prospectively only. Any such limitation does
not alter Client's obligation to reimburse contractor for
only reasonable expenses incurred.

3.3 Contractor shall provide Client with an invoice for
said fees and reimbursable expenses at the end of each
calendar month. Said invoices shall be due and
payable on or before the tenth (10th) day of the
following month.

**4. TERMINATION
OF AGREEMENT**

This Agreement will terminate on
__________________, 19______, unless terminated
sooner as provided herein. Despite the foregoing,
either Client or Contractor can terminate this
Agreement upon thirty (30) days written notice of
termination to the other party. Further, if any of the
following occur, this Agreement shall terminate
immediately, upon notice to the other party.

4.1 Election of Client to sell its entire business, or
cease doing business;

4.2 The entry against Contractor by a court of
competent jurisdiction of a decree or order constituting
an adjudication of bankruptcy under federal
bankruptcy law or of insolvency under any state
insolvency law; the initiation of proceedings by
Contractor seeking reorganization or adjudication as a
voluntary bankrupt; consent to the filing of a
reorganization, insolvency, or bankruptcy proceeding or
the appointment of a conservatory, custodian, receiver,
liquidator, trustee, or assignee in bankruptcy or
insolvency under any applicable federal or state law;
an assignment by Contractor for the benefit of
creditors, or an admission in writing by Contractor of
inability to pay its debts generally as they become due.

4.3 Conviction of Contractor or any of its employees of any felony, or any misdemeanor involving moral turpitude;

4.4 Failure of Contractor to diligently or properly perform its duties under this Agreement; and

4.5 Failure of Client to pay Contractor for properly invoiced amounts within ten (10) days of written notice given to Client by Contractor that said invoice is past due.

5. STATUS OF CONTRACTOR AS AN INDEPENDENT CONTRACTOR

This Agreement does not constitute a hiring by either party. It is the parties' intention that Contractor shall be an independent contractor and not Client's employee for all purposes, including, but not limited to, the application of the Federal Insurance Contribution Act, the Social Security Act, the Federal Unemployment Tax Act, the provisions of the Internal Revenue Code, the California Revenue & Taxation Code relating to income tax withholding at the source of income, California Workers'Compensation Act and the California Unemployment Insurance Code. Contractor shall retain sole and absolute discretion in the judgment of the manner and means of carrying out Contractor's activities and responsibilities hereunder. Contractor shall be entitled to engage in any activities which are not expressly prohibited by this Agreement. This Agreement shall not be construed as a partnership and Client shall not be liable for any obligation incurred by Contractor.

6. HOURS WORKED

Contractor shall perform at a minimum, __________ hours per month. The scheduling of hours worked shall be within the sole discretion of Contractor.

7. NOTICE CONCERNING WITHHOLDING OF TAXES

7.1 Since Contractor is a corporation, it will not be required to pay federal self-employment taxes upon amounts received by it under this Agreement. However, Contractor recognizes and understands that it will be required to file a corporate tax return and to pay tax upon its corporate taxable income in accordance with all the provisions of applicable state and federal law. In addition, Contractor recognizes and understands that it is required to withhold taxes from wages paid to its employees in accordance with

the provisions of state and federal law. Contractor
hereby promises and agrees to indemnify Client for
any damages or expenses, including attorneys' fees and
legal expenses, incurred by Client as a result of
Contractor's failure to make such required payments.
Contractor further recognizes and understands that it
is required to provide adequate workers' compensation
insurance for its employees or to adequately self-insure
against such risks in accordance with California law
and will comply with said laws. Contractor further
recognizes and understands that it is required by
California law to make contributions to the State of
California pursuant to the California Unemployment
Insurance Code and will comply with said laws.

7.2 Contractor will provide Client, on a monthly basis,
with proof that it has withheld all required amounts
from the compensation paid to its employees and paid
said amounts, along with other required additional
amounts, to the state and federal authorities.
Contractor will further provide Client on a monthly
basis with proof that it has in force an adequate
workers' compensation insurance policy or is properly
self-insured.

**8. ATTORNEYS'
FEES AND LITI-
GATION COST**

If any legal action or any arbitration or other
proceeding is brought for the enforcement of this
Agreement, because of an alleged dispute, breach or
default in connection with any of the provisions of
this Agreement, the successful or prevailing party shall
be entitled to recover reasonable attorneys' fees and
other costs incurred in that action or proceeding, in
addition to any other relief to which that party would
be entitled.

**9. ENTIRE
AGREEMENT**

This Agreement constitutes the entire agreement of the
parties and supersedes all prior agreements,
understandings and negotiations, whether written or
oral, between the parties. This Agreement may not be
changed orally but only by an agreement in writing
signed by both parties which is expressly stated to be
an amendment hereto.

10. PROVISIONS SEVERABLE

In case any one or more provisions of this Agreement, shall be invalid, illegal or unenforceable in any respect, the validity, legality and enforceability of the remaining provisions contained herein shall not be in any way affected or impaired thereby.

11. HEADINGS

Headings set forth herein are for the convenience of the parties only and are not a part of the Agreement.

12. INTERPRETATION

This Agreement shall be construed and interpreted under the laws of the State of ____________________.

[Continued on Next Page]

13. NOTICES All notices herein shall be in writing. Notices may be delivered personally, or by mail, postage prepaid, to the respective addresses noted below. Any notice sent by mail, postage prepaid, will be deemed received three days after it is mailed. Each party may designate a new address for purposes of this Agreement by notice to the other party.

IN WITNESS WHEREOF, the parties hereto have executed this Agreement at ________________________________, to be effective as of

________________________________, 19______.

"CLIENT"

X ________________________________

Address: ________________________________

____

By: ________________________________

Its: ________________________________

"CONTRACTOR"

X ________________________________

Address: ________________________________

____

By: ________________________________

Its: ________________________________

CAUTION: This form is an illustration only. It has not been prepared with any specific factual situation in mind. It may not be suitable for your particular use. No warranty is made as to its fitness for any particular purpose or use. No other warranty is given. Before using this agreement, or any part of it, you should consult with your attorney.

This page is intentionally left blank.

AGREEMENT No. 4

SUBHAULING SERVICES

Comment:

This independent contractor agreement deals with delivery services, specifically "subhauling". Section 3 of the agreement on comprehensive general liability insurance and comprehensive automobile/vehicle liability insurance is a good idea. If an insurance clause is called for, you should state in detail what insurance is required. With respect to paragraph 2(k), it is not a good idea to require the worker to wear uniforms of the employer.

AGREEMENT

THIS INDEPENDENT CONTRACTING SUBHAULER'S AGREEMENT, made and

entered into between ____,

a _________________________________ Corporation with offices located at

____ (hereinafter referred to as

"CLIENT"), and ____, an independent

contracting subhauler (hereinafter referred to as "CONTRACTING

SUBHAULER");

WITNESSETH THAT:

WHEREAS, _________________________________ is a common carrier by motor

vehicle engaged in the contract pick-up and delivery business for ABC Air Express,

and for such purposes requires pick-up and delivery service in the city and county

of _________________________________ and its vicinity, pursuant to authority

granted by the _________________________________ Public Utilities Commission;

and,

WHEREAS, the CONTRACTING SUBHAULER above-named, engaged in the

business of public transportation, is willing and able to render pick-up and

delivery service upon such terms and conditions herein provided for, and as may

22

from time to time be requested by ________________________________, and in
return for said services provided hereunder.

NOW, THEREFORE, it is mutually covenanted and agreed by and between the
parties hereto as follows:

SECTION 1. TERM

This Independent Contracting Subhauler's Agreement shall become effective immediately upon the execution hereof, on the date and year set forth hereinbelow, and shall remain in effect for a period of one (1) year; thereafter it shall be effective from year to year unless terminated earlier by either party hereto as hereinafter provided.

SECTION 2. CONTRACTING SUBHAULER'S SERVICES

During the existence of this Independent Contracting Subhauler's Agreement, CLIENT may request pick-up, transport and delivery services from the CONTRACTING SUBHAULER at the rates provided by Section 2 hereof, and the CONTRACTING SUBHAULER may accept such request and perform such services. The CONTRACTING SUBHAULER, upon accepting such request, shall perform those services necessary or incidental to the performance of the contracts between CLIENT and ABC Air Express, or as said contracts may from time to time be altered, changed, or modified, and the shipper, including the following:

(a) The CONTRACTING SUBHAULER shall, as an independent contractor, pick-up, transport and deliver shipments from the airport(s) serving the aforementioned city and county to consignees, and from consigners to said airport(s) as from time to time requested by CLIENT in connection with its ABC Air Express freight business.

(b) The CONTRACTING SUBHAULER shall provide and operate, or cause to be operated, for such pick-up, transport and delivery service such vehicle(s) and equipment necessary to the hauling of shipments which are in the independent and exclusive charge and control of the CONTRACTING SUBHAULER, and the CONTRACTING SUBHAULER shall furnish or cause to be furnished, an adequate supply of labor in connection therewith.

(c) The geographical scope of the pick-up and delivery services provided by the CONTRACTING SUBHAULER for CLIENT shall be limited solely to, and shall not exceed the air freight terminal area of the airport city shown in the Agreement set forth in the tariff identified as ABC Air Express. Air Freight Pick-up and Delivery tariff, No. ______________, or as that tariff may be revised or reissued from time to time by ABC Air Express. No other operating authority whatsoever is either expressed or implied by this Independent Contracting Subhauler's Agreement.

(d) When so requested by CLIENT, the CONTRACTING SUBHAULER shall load and unload shipments, or cause such shipments to be loaded and unloaded, and shall furnish or cause to be furnished whatever labor is required for such loading and unloading.

(e) When so requested by CLIENT, the CONTRACTING SUBHAULER shall perform such other transportation or incidental or supplemental accessorial services as may be necessary and consistent with the pick-up, transportation and delivery contracts between CLIENT and ABC Air Express and their customers, and to complete the several portions and the whole of the services accepted and undertaken by the CONTRACTING SUBHAULER under this Independent Contracting Subhauler's Agreement, in such timely and expeditious manner as to enable CLIENT and ABC Air Express to timely perform and complete their obligations to each other and the shipper.

(f) The CONTRACTING SUBHAULER shall be solely responsible for and shall pay all expenses incurred in the acceptance and performance of his assumed obligations under this Independent Contracting Subhauler's Agreement and the services listed in this Section 2 hereof. In connection therewith, the CONTRACTING SUBHAULER

shall, in the sound and prudent exercise of his
independent, sole and exclusive judgment, and subject
only to such actual restrictions hereafter specified;

  (1) Hire, compensate, supervise and
control whatever labor he deems necessary for the
performance of such services, including a fully trained
relief or substitute driver, provided that CLIENT is
given sufficient evidence to show that such driver
meets all applicable governmental requirements;

  (2) Maintain whatever records he desires,
in addition to those required by law, governmental
rules, regulations and orders, or that may be from time
to time promulgated and required to be maintained by
ABC Air Express and over which CLIENT has no
control whatsoever;

  (3) Select whatever routes that are to be
traversed in performing his contract:

  (4) Garage, store, maintain and repair
his vehicle(s) and equipment, and purchase fuel and oil
at places of his own choosing alone;

  (5) Determine how many hours he will
work a day within the limitations imposed by law,
what days he will work, vacations and so on; and,

  (6) At any time whatsoever reject any
requested assignments at his sole and absolute
discretion.

  (g) The CONTRACTING SUBHAULER'S
vehicle(s) and equipment shall be painted to such
specifications as shall from time to time be
promulgated and required by ABC Air Express, and to
be repainted thereafter whenever the CONTRACTING
SUBHAULER determines that a reasonable need
thereof exists, which shall be determined solely in the
absolute and exclusive judgment of the
CONTRACTING SUBHAULER.

(h) The CONTRACTING SUBHAULER shall
provide whatever terminal handling assistance as may
be necessary in the loading and unloading of trucks,
and for the sorting and labeling of shipments at the
ABC Air Express terminal.

(i) The CONTRACTING SUBHAULER shall
complete all deliveries of ABC Air Express shipments
by no later than 12 o'clock noon.

(j) The CONTRACTING SUBHAULER shall
provide, or cause to be provided, pick-up service of
ABC Air Express shipments within ninety (90) minutes
of pick-up call, and shall provide delivery service of
all shipments within two (2) hours of recovery thereof.

(k) The CONTRACTING SUBHAULER, and
his employees and agents providing service provided
hereunder, shall wear ABC Air Express uniforms while
engaged in the performance of his obligations under
this Independent Contracting Subhauler's Agreement,
and they shall present a neat and well-groomed
appearance.

(l) The CONTRACTING SUBHAULER shall
provide all necessary airport terminal coverage to
recover or tender airline consolidations as needed.

(m) The CONTRACTING SUBHAULER and
his employees, assume sole and exclusive responsibility
for compliance with all economic, operational, safety,
insurance, and any other requirements imposed by
Federal, State, County, Municipal, or other law or
regulatory body, relating to the surface operations
hereunder; and the CONTRACTING SUBHAULER
agrees to reimburse any and all like costs incurred by
CLIENT, including but not limited to the amounts of
fines or penalties and costs of counsel, arising from
any assertion or finding of lack of compliance with the
aforesaid laws and/or regulations in respect to the
surface operations hereunder. The CONTRACTING
SUBHAULER agrees that CLIENT may deduct the
amounts due it hereunder from any monies otherwise
owed to CONTRACTING SUBHAULER, in full or
partial payment of his obligation to reimburse to
CLIENT its costs as hereinabove defined.

(n) The CONTRACTING SUBHAULER shall
maintain his vehicle(s) and equipment in good repair
and will immediately take whatever action is necessary,
including periodic safety inspections, to assure that
said vehicle(s) and equipment meet the requirements of
the State of ______________________ and of the
United States Department of Transportation
prerequisite to proper and safe operation.

The CONTRACTING SUBHAULER
warrants that his vehicle(s) and equipment is in good
repair and complies with all safety requirements of the
State of ______________ and of the United States
Department of Transportation.

(o) The CONTRACTING SUBHAULER shall
pay all costs of operation of his vehicle(s) and
equipment, including resident vehicle license,
insurance, property taxes, fuel, oil, tires, repairs and
tolls and all other costs incident to ownership and
operation of the vehicle(s) and equipment, and the
CONTRACTING SUBHAULER shall pay all license,
permit, and other fees and costs.

(p) When required, the CONTRACTING
SUBHAULER shall provide and maintain at his
expense a Corporate, Subhauler's, or other similar type
of Surety Bond in a minimum amount satisfactory to
CLIENT conditional for the full and faithful
performance of this Independent Contracting
Subhauler's Agreement and payment for labor, services,
and other expenses, materials and costs covered hereby.

(q) Nothing herein shall prevent the
CONTRACTING SUBHAULER from performing any
transportation service for any other person, firm,
corporation, or company whatsoever, provided that the
CONTRACTING SUBHAULER shall not engage in
other transportation service of air freight goods
without giving prior written notice of CONTRACTING
SUBHAULER'S intention thereof to CLIENT.

**SECTION 3.
WORKERS' COM-
PENSATION AND
PUBLIC LIABILITY
AND PROPERTY
DAMAGE INSUR-
ANCE COVERAGE:**

**CONTRACTING
SUBHAULER'S
AND CLIENT'S
RESPECTIVE
LIABILITIES.**

(a) The CONTRACTING SUBHAULER shall carry and maintain at the CONTRACTING SUBHAULER'S sole expense, insurance in such form(s) and amount(s) as CLIENT may require, but in no event less than the minimum amounts of insurance coverage hereinafter set forth, and the CONTRACTING SUBHAULER shall furnish to CLIENT a certificate from any and all insurance carriers showing the date(s) of expiration of any and all policies, limits of liability thereunder, and providing that said insurance will not be canceled or changed until or unless at least thirty (30) days prior written notice given to CLIENT, and then only canceled or changed with written consent of CLIENT.

(b) The CONTRACTING SUBHAULER shall name CLIENT as an additional insured on any and all policy or policies of insurance provided hereunder.

(c) In the event that the CONTRACTING SUBHAULER shall fail to furnish and maintain such insurance coverage provided for hereunder, CLIENT shall have the right to take out and hereunder, CLIENT shall have the right to take out and maintain the said insurance for and in the name of the CONTRACTING SUBHAULER, and the CONTRACTING SUBHAULER shall pay the cost thereof and shall furnish all necessary information to make effective and maintain such insurance.

(d) The CONTRACTING SUBHAULER shall carry and maintain at least the following minimum limits of insurance coverage:

(1) Comprehensive General Liability Insurance

Personal Injury:
Minimum $100,000 each person
Minimum $300,000 each occurrence
Minimum $300,000 aggregate

Property Damage:
Minimum $100,000 each accident
Minimum $100,00 aggregate

(2) Comprehensive Automobile/
Vehicle Liability

Personal Injury:
Minimum $100,000 each person
Minimum $300,000 each occurrence
Minimum $300,000 aggregate

Property Damage:
Minimum $100,000 each accident

(e) The CONTRACTING SUBHAULER may purchase such insurance as he may determine in addition to the minimum insurance required to be maintained by him under this Section, including workers' compensation insurance for the CONTRACTING SUBHAULER'S own business and employees; and, in the event that the CONTRACTING SUBHAULER shall purchase and maintain such workers' compensation insurance coverage for his own business and employees, then the CONTRACTING SUBHAULER shall carry and maintain at least the following minimum limits of said workers' compensation coverage: $100,000 minimum (one or more employees).

(f) The CONTRACTING SUBHAULER hereby agrees to assume the entire responsibility in and for any and all damage or injury of any kind or nature whatsoever to all persons, whether employees or otherwise, and to all property of any and every kind and description whatsoever, and wherever situated, growing out of or resulting from the execution of work provided for in this Independent Contracting Subhauler's Agreement or occurring in connection therein, not otherwise covered by insurance, and agrees to indemnify and save harmless CLIENT, its agents, servants, and employees, from and against any and all loss, expense, including reasonable attorneys' fees and court-related costs, damage or injury growing out of or resulting from or occurring in connection with the execution of the work herein provided for, or the performance or lack of performance thereof, or occurring in connection with or resulting from the use by the CONTRACTING SUBHAULER, his agents or employees, of any and all vehicles, materials, tools,

implements, appliances, scaffolding, ways or hoists,
elevators, works or machinery, or other property of
either CONTRACTING SUBHAULER or CLIENT,
whether the same arise under the common law or
otherwise and not covered under the so-called
_______________________ Workers' Compensation law.

(g) Nothing hereunder shall prevent CLIENT,
as a cost of its doing business, from carrying and
maintaining any and all types of insurance coverage in
its own name and at its own expense, which insurance
coverage shall be excess insurance coverage over and
above any valid and collectible insurance coverage
carried by the CONTRACTING SUBHAULER.

(h) Unless otherwise agreed upon by and
between the parties hereto, CLIENT shall carry and
maintain as a cost of doing business and at its own
cost, the following minimum limits of insurance
coverage:

(1) Workers' Compensation
(one or more employees)

Minimum $100,00

Nothing hereunder shall prevent the
CONTRACTING SUBHAULER however, from
carrying and maintaining said workers' compensation
insurance in his own name and at his own expense
alone, for the CONTRACTING SUBHAULER'S own
business and employees; and, in that event, the
CONTRACTING SUBHAULER shall furnish to
CLIENT a certificate from any and all insurance
carriers providing such workers' compensation
insurance coverage hereunder, showing the date(s) of
expiration of any and all such policies, the limits of
liability thereunder, and providing that said insurance
will not be canceled or changed until and unless at
least thirty (30) days prior notice given in writing to
CLIENT, and then only canceled or changed with the
written consent of CLIENT.

SECTION 4.
SOCIAL SECURITY
AND
WITHHOLDING
TAXES

It is the intention of the parties that the CONTRACTING SUBHAULER shall control the means of and be an independent contractor in performing services under this Independent Contracting Subhauler's Agreement. The CONTRACTING SUBHAULER shall be solely responsible for the payment,

     (a) of Social Security, Withholding, and all other taxes imposed by reason of his employment of persons to perform labor, and,

     (b) of all taxes which result from the payment of funds for services rendered to the CONTRACTING SUBHAULER pursuant to this Independent Contracting Subhauler's Agreement. Nothing hereunder, however, shall prevent CLIENT from voluntarily assisting the CONTRACTING SUBHAULER in maintaining his records and books of such Social Security and Withholding taxes, including the assistance in computing the proper amounts of such taxes and the actual withholding of such taxes when requested by the CONTRACTING SUBHAULER.

SECTION 5.
CLAIMS BY
CONSIGNORS AND
CONSIGNEES

ABC Air Express will assume the liability of CLIENT and the CONTRACTING SUBHAULER for all risks of loss or damage to ABC Air Express' shipments, including the liability arising under the CONTRACTING SUBHAULER'S agreement to indemnify ABC Air Express for any such loss or damage, while such shipments are in the custody or control of the CONTRACTING SUBHAULER. In consideration of such assumption of liability, CLIENT and CONTRACTING SUBHAULER hereby authorize ABC Air Express to deduct and retain from compensation due to CLIENT and CONTRACTING SUBHAULER, for cartage services rendered, an amount equal to two percent (2%) of such compensation.

SECTION 6.
COMPENSATION
TO CONTRACTING
SUBHAULER

CLIENT shall pay for the requested service furnished to it in accordance with this contract at the rates set forth in Schedule A attached hereto, less ___________% deducted therefrom, to cover services provided, and such costs incurred, by CLIENT.

SECTION 7.
TERMINATION

This Independent Contracting Subhauler's Agreement may be terminated for any reason by either party hereto.

SECTION 8.
MISCELLANEOUS
PROVISIONS

(a) No assignment of this contract shall be made by either party without the written consent of the other.

(b) In furnishing the services referred to above, the CONTRACTING SUBHAULER may be required to collect upon delivery such charges as directed by CLIENT and ABC Air Express which will consist of shipper's C.O.D. charges or transportation charges, or both. Such charges are to be remitted to ABC Air Express in accordance with procedures outlined from time to time by said ABC Air Express.

(c) CLIENT shall maintain a Subhauler's Surety Bond in the minimum amount of $10,000.00.

(d) If any action at law or in equity is necessary to enforce or interpret the terms of this Independent Contracting Subhauler's Agreement, the prevailing party shall be entitled to reasonable attorneys' fees and costs in addition to any other relief to which he may be entitled.

(e) This Independent Contracting Subhauler's Agreement supercedes any and all other agreements, either oral or in writing, between the parties hereto and contains all of the covenants and agreements between the parties with respect to their rights, duties, and obligations toward each other. Each party to this Independent Contracting Subhauler's Agreement acknowledges that no representations, inducements, promises, or agreements orally or otherwise, have been made by any party, or anyone acting on behalf of any party, which are not embodied herein, and that no other agreement, statement, or promise not contained in this Independent Contracting Subhauler's Agreement shall be valid or binding. Any modification of this contract will be effective only if it is in writing signed by the party to be charged.

(f) If any provision in this Independent Contracting Subhauler's Agreement is held by a court of competent jurisdiction to be invalid, void, or unenforceable, the remaining provisions shall nevertheless continue in full force without being impaired or invalidated in any way.

(g) This Independent Contracting Subhauler's Agreement shall be governed by and construed in accordance with the laws of the State of ___________.

(h) Should there be a dispute concerning the amount to pay the CONTRACTING SUBHAULER pursuant to the services rendered to CLIENT, the CONTRACTING SUBHAULER and CLIENT agree that the disputed matter shall be submitted to arbitration pursuant to the American Arbitration Association, and that any decision rendered thereby shall be binding upon the parties to this contract in any court of law or otherwise.

(i) This Independent Contracting Subhauler's Agreement has been executed in duplicate, one (1) copy to be retained by CLIENT, and another copy by the CONTRACTING SUBHAULER, receipt of a copy of which is hereby acknowledged. Both parties agree that they have received a copy of this Independent Contracting Subhauler's Agreement, have read and understood each and every term hereof, and that they have conferred with counsel as to any matters to which they have any questions as to the effect and meaning

[Continued on Next Page]

of any such provisions of this Independent Contracting Subhauler's Agreement, and that they fully understand the same.

IN WITNESS WHEREOF, the parties have set their hands and seals this

_______________________________day of_____________________, 19_________.

    "CONTRACTING SUBHAULER"   **"CLIENT"**

Calt#_________________________   Calt#_________________________

By:___________________________   By:___________________________

Its: _______________________   Its:___________________________

Witness:______________________   Witness:______________________

SCHEDULE A

RATES AND CHARGES TO BE PAID BY CLIENT FOR:

1. <u>Pickup, Delivery, Transfer and Reforwarding Service:</u>

TYPE OF SERVICE	CWT RATE	MINIMUM CHARGE
Local Pickup & Delivery - Per Stop	$________	$________
	$________	$________
	$________	$________
<u>Inter-Airline Transfer at Airport</u>	$________	$________
<u>Affixing Postage - Per Package</u>	-------	$________
Other than Local Pickup & Delivery (List Points)		
	$________	$________
	$________	$________
	$________	$________
	$________	$________
	$________	$________
	$________	$________
	$________	$________

Local Pickup & Delivery Area Includes:

2. <u>Collecting and Remitting C.O.D. Amounts</u> (Applies only to shipper's C.O.D. amounts. Does not apply to the collection and remittance of Client's freight charges.)

AMOUNT OF C.O.D.	CHARGE	AMOUNT OF C.O.D	CHARGE
Not Over $100	$______	From $500.01 to $ 600	$______
From $100.01 to 200	______	" 600.01 to 700	______
" 200.01 to 300	______	" 700.01 to 800	______
" 300.01 to 400	______	" 800.01 to 900	______
" 400.01 to 500	______	" 900.01 to 1000	______

Add $.10 for each $100 or fraction thereof above $1,000.

    By Mutual agreement between Client and the Contractor, this Schedule is hereby made part of the Agreement between the two parties dated ____________.

    The effective date of this Schedule is ____________________.

CONTRACTOR   CLIENT

BY________________________________   BY ________________________________
        President

AGREEMENT No. 5

HOME MAINTENANCE SERVICES

Comment:

This agreement deals with the household cleaning business. Paragraph 5 may not contain the best wording because it states that the client or the owner may be providing instrumentalities. True independent contractors generally provide their own instrumentalities. A few words of caution are in order: Workers in this area are generally considered to be low skill. In my view, there is a high risk that household cleaning crews will be found to be employees of the household cleaning company no matter what the agreement says.

AGREEMENT

This Independent contractor agreement ("Agreement") is entered into by and between _______________________ ("OWNER") and the undersigned independent contractor ("CONTRACTOR") on the date and at the place set forth below:

WHEREAS, the OWNER is in the business of providing house cleaning and other services for homeowners and others and has and develops or obtains clients desiring such services; and

WHEREAS, the CONTRACTOR is ready, willing and able to provide house cleaning and other services for homeowners and others;

NOW THEREFORE, it is mutually agreed as follows:

1. CLIENTS	The OWNER will supply the CONTRACTOR with clients of the OWNER to service;
2. STANDARD OF SERVICE	The CONTRACTOR shall work diligently and use best efforts to service clients of the OWNER as necessary in providing house cleaning and other services thereto;
3. PAYMENT FOR SERVICES	The OWNER will pay an agreed upon amount to the CONTRACTOR for services rendered to each client of the OWNER;
4. HOURS WORKED	The CONTRACTOR'S days and hours of performing such services are variable from day to day and week to week and are mutually determined and agreed upon based upon the requirements and needs of the clients and the availability of the CONTRACTOR to render such services;

5. CLEANING MATERIALS, EQUIPMENT AND TOOLS	The OWNER, clients of the OWNER or the CONTRACTOR shall provide all cleaning materials, equipment and tools necessary, as agreed upon by the parties either orally or in writing;
6. LIABILITY FOR ACTS OF CONTRACTOR **CONTRACTOR'S INDEPENDENT STATUS**	The OWNER shall not be liable for any negligent, reckless or intentional acts or omissions of the CONTRACTOR, nor shall the CONTRACTOR bind or attempt to bind the OWNER in any manner; and nothing herein shall be construed as creating the relationship of employer and employee between the parties, but rather, the CONTRACTOR shall, at all times, be deemed to be an independent contractor and free of any control by the OWNER in selecting the time or method of work or transportation used;
7. CLAIM BY A CLIENT FOR LOSS, INJURY OR DAMAGE	The OWNER may withhold $50.00 from any sum or sums due or accruing to the CONTRACTOR for services performed in the event of any claim by a client of the OWNER due to loss, injury or damage as the result of services rendered or provided by the CONTRACTOR, to be applied to the cost thereof, with any excess amount withheld paid to the CONTRACTOR after the cost of such loss, injury or damage is ascertained; any damage amount between $50 and $250 will be shared equally between the OWNER and contractor.
8. THIRD PARTY CLAIM FOR LOSS, INJURY OR DAMAGE	The CONTRACTOR shall indemnify and hold the OWNER harmless from all losses, injuries or damages caused by the negligent, reckless or intentional acts or omissions of the CONTRACTOR in rendering services pursuant to this Agreement, including payment of reasonable attorneys' fees and costs in the defense of any claim made by a third person incident to such negligent, reckless or intentional act or omission;
9. MANUALS, BOOKLETS AND MATERIALS	All manuals, booklets and materials of the OWNER, and all records of the OWNER'S clients, route books, and any other records, books or other materials relating in any manner whatsoever to clients of the OWNER, whether prepared by the OWNER or otherwise coming into the OWNER'S possession, shall be the exclusive property of the OWNER regardless of who actually purchased or created the original books, records or

materials, and all such books, records and material shall be immediately returned or delivered over to the OWNER by the CONTRACTOR upon demand therefore;

10. IDENTITY AND/OR SOLICITATION OF OWNER'S CLIENTS

The CONTRACTOR shall not make known to any person, firm, corporation, or business the names and/or addresses of any of the clients of the OWNER, or any other information pertaining to them, or call on, solicit, take away, or accept, directly or indirectly, business from such clients, or attempt to call on, solicit, take away, or accept, directly or indirectly, business from such clients, with whom the CONTRACTOR becomes acquainted during and/or as a result of services rendered pursuant to this Agreement, whether for the CONTRACTOR'S benefit, or potential benefit, or the benefit, or potential benefit, of any other person, firm, corporation, or business during the term of this Agreement, or for a period of eighteen (18) months after the termination of this Agreement;

11. CONTRACTOR SHALL NOT PROVIDE CLIENT WITH THE MEANS TO CONTACT HIM EXCEPT THROUGH OWNER

The CONTRACTOR shall not provide the clients of the OWNER with the CONTRACTOR'S home telephone number or any other information enabling such clients to contact the CONTRACTOR other than through the OWNER.

[Continued on Next Page]

**12. TERM OF
AGREEMENT**
The term of this Agreement shall commence on its execution and shall terminate upon seven (7) days notice, oral or written, from one party to the other.

Executed on _____________________________, 19________, at

_______________________, State of _____________________________.

By: _______________________

      (CONTRACTOR)

By: _______________________

      (OWNER)

This page is intentionally left blank.

Comment:

This independent contractor agreement deals with management services, specifically, the management of a concrete plant. Article 9 on indemnification is worthy of note. If the independent contractor gets the employer in trouble, the independent contractor agrees to make the employer whole and to pay his costs of defense. Not a bad idea if you have an independent contractor of some substance that can actually carry through on that contract promise. However, in many cases insurance may be the preferred way to take care of the indemnification problem.

AGREEMENT

ARTICLE 1 **PARTIES**	An Agreement was made on ________________________, 19______, between ____, hereinafter referred to as "CONTRACTOR," and ____, a _______________________ corporation, hereinafter referred to as "CORPORATION." Said Agreement is now restated and clarified effective __________________, 19______.
ARTICLE 2 **BASIC AGREEMENT**	CONTRACTOR hereby proposes to furnish to CORPORATION all services to perform the work of managing the concrete plant facility known as ___________________________________, located at ____, for the price stated and in accordance with the specifications set out below. On properly executed acceptance of this contract, CORPORATION agrees to purchase such services, subject to all other terms and conditions of this restatement of agreement.
ARTICLE 3 **MANAGEMENT FEE**	The price for furnishing the service of managing the concrete plant facility will be thirty-three and one-third percent (33-1/3%) of the profits from the operation of the concrete plant facility or \$2,000.00 per month, whichever sum is greater. CORPORATION will pay CONTRACTOR monthly for work done to date.
ARTICLE 4 **DEFINITION OF PROFITS**	The term "profits" as used herein shall be determined by deducting from the income of the concrete plant facility all of the direct operating expenses and overhead expenses, including but not limited to, the cost of utilities, rent of real estate, rental of equipment, property taxes, depreciation, interest on funds borrowed for the concrete plant facility, administrative expenses charged by CORPORATION, and all other overhead expenses, it being the intent to determine the profits in accordance with generally accepted accounting principles.

ARTICLE 5

**PROFESSIONAL
MANAGEMENT
STANDARDS**

CONTRACTOR agrees to operate and maintain the
concrete plant facility, including all buildings,
inventory, equipment, real property, supplies and
materials used in connection therewith, in a manner
calculated to enhance the reputation of the concrete
plant facility with its customers. CONTRACTOR
agrees to use its best efforts in managing said facility
in order to provide the maximum economic return
consistent with professional management standards.

CONTRACTOR shall have full power and authority to
manage the facility and shall be responsible for
directing its supervisors and employees as to the
manner and means of accomplishing the work required
to be performed.

CONTRACTOR agrees, within a reasonable time after
the parties hereto have executed this Agreement, to
inform itself with respect to the inventory, office
equipment and supplies, machinery, and the mechanical
and physical systems operating within the facility.
CONTRACTOR will also provide CORPORATION
with inventories of all equipment, supplies, personal
property and readily movable fixtures attached to the
real property and buildings at the facility.

ARTICLE 6

**PRECAUTIONARY
PROCEDURES**

CONTRACTOR will take all required necessary and
customary precautions in introducing and maintaining
safety measures against all hazards likely to be
connected with the types of work involved hereunder.
Protective arrangements will be taken in all instances
to prevent work operations from damaging the premises
or other work or operations in any way.

ARTICLE 7

**COMPLIANCE
WITH GOVERN-
MENTAL REGU-
LATIONS**

CONTRACTOR shall take all required action to
comply promptly with all Federal, State, County and
Municipal rules, regulations and orders, provided,
however, that if CORPORATION is contesting or has
affirmed its intention to contest any such rule,
regulation or order, CONTRACTOR shall not take any
action under this paragraph. CONTRACTOR shall,
within 24 hours of receipt of any Federal, State,
County or Municipal rule, regulation or order, notify
CORPORATION in writing of its receipt of such
order, rule or regulation.

ARTICLE 8

**INSURANCE
COVERAGE**

CORPORATION shall maintain in full force and effect all policies of insurance now existing in connection with the concrete plant facility, the buildings and equipment thereon, and the inventory, including but not limited to public liability insurance, product liability insurance, property damage and personal injury insurance, and fire and theft coverage. CORPORATION may designate CONTRACTOR as a co-insured on one or more of said policies of insurance if in the judgment of its insurance agent such designation is warranted. The cost of any such insurance coverage shall be an operating expense of the concrete plant facility.

ARTICLE 9

**INDEMNIFI-
CATION**

CONTRACTOR assumes all liability for and agrees to defend, indemnify and hold CORPORATION, its employees and its agents or subsidiaries, harmless from all loss, damage, cost and expense, including all attorneys' fees incurred by CORPORATION arising from or in any way connected with the CONTRACTOR'S operations or the operations of any subcontractor, agent, servant or employee of the CONTRACTOR, including without limitation, bodily injury, sickness and/or disease, including death at any time, resulting from such bodily injury, sickness and/or disease, sustained by any person while in, on or about the concrete plant facility, or in connection with the operation of any personal vehicle of the CONTRACTOR or of its subcontractors, agents, servants or employees, or in connection with the rental or use of any item or items of inventory if or when such injury, sickness, disease and/or death arose out of or was in any way connected with the operation or the performance of or failure to perform any duty, obligation, or activity in connection with the management of said facility.

ARTICLE 10

**COLLECTION OF
RECEIPTS**

CONTRACTOR shall use its best efforts to collect all receipts due from the sale of concrete materials and rentals of equipment used to transport, mix, carry and unload said concrete materials, together with sales of other inventory items as may be authorized by CORPORATION. CONTRACTOR and any agents, servants or employees of CONTRACTOR who handle or who are responsible for the handling of CORPORATION'S monies shall be bonded by a fidelity acceptable to both CONTRACTOR and CORPORATION, indemnifying CORPORATION as obligee against loss, theft, embezzlement, or other fraudulent acts on the part of CONTRACTOR or its agents, servants, employees or subcontractors, and such bonding expense shall be a part of the operating expense of the concrete plant facility.

ARTICLE 11

RECEIPTS

CONTRACTOR shall use and maintain a checking account acceptable to CORPORATION and to CONTRACTOR for the deposit of all receipts from the sales of concrete materials and the rental of inventory in connection with the operation of the facility.

ARTICLE 12

PAYMENT OF EXPENDITURES

CORPORATION, with the assistance of CONTRACTOR, shall cause to be disbursed regularly and punctually the recurring operating expenses of the facility, including all payments for assessments and encumbrances on the property.

CONTRACTOR shall institute and maintain a control system to insure the authenticity of bills received. CONTRACTOR shall authorize not more than two of its employees to be responsible for the approval of all purchase requests and hiring of services for the facility, provided, however, that CORPORATION may, from time to time, submit to CONTRACTOR a list of suppliers and services for the facility.

ARTICLE 13

FINANCIAL RECORDS AND REPORTS

CONTRACTOR shall keep accurate and complete records in accordance with the accounting standards and procedures presently utilized by CORPORATION in other rental facilities so that all receipts from the sale of concrete materials and the rental of inventory may be ascertained on a daily basis by CORPORATION. CORPORATION shall have the right at any reasonable time to inspect any such record of CONTRACTOR in order to verify the financial reports of CONTRACTOR, including but not limited to all checks, bills, vouchers, invoices, statements, cash receipts, correspondence, and all other records in connection with the management of the facility. CORPORATION shall further have the right to cause an audit to be made of all account books and records connected with the management of the concrete plant facility.

CONTRACTOR shall prepare a monthly statement
showing in detail all of the receipts and disbursements
from the preceding month, and shall prepare a
quarterly summary of receipts and disbursements, such
monthly and quarterly reports to be submitted to
CORPORATION within 15 days after the close of the
month or quarter, whichever is appropriate.

ARTICLE 14

**PURCHASE OF
INTEREST**

On _______________________, 19_______,
CONTRACTOR purchased a one-third interest in the
equipment, inventory and accounts receivable of the
concrete plant facility. On that date, CONTRACTOR
also executed a promissory note in favor of
CORPORATION in the principal amount of $10,000.00,
payable in five (5) years from ___________________,
19_______, with interest at 8% per year, with monthly
payments of $_______________ or more until paid in
full, with no prepayment penalty. Effective
_______________________, 19_______.
CONTRACTOR also assumed, became liable for, has
paid, and will continue to pay to CORPORATION
monthly that sum which represents one-third (1/3) of
the total monthly payments due on the encumbrances
of the equipment and other assets of the concrete plant
facility that now exist and that may hereafter be
acquired.

Either CONTRACTOR or CORPORATION may cause
the monthly payments due on said promissory note at
the rate of $_______________ per month to be paid
from the CONTRACTOR'S share of the profits from
the operation of the concrete plant facility, as the term
"profits" is defined in ARTICLE 4 of this Agreement.

Should monthly profits from the operation of the
concrete plant facility fall below the monthly
operating expenses thereof for a period of six (6)
consecutive months, CORPORATION shall have the
exclusive option to purchase CONTRACTOR'S interest
in the concrete plant facility. Said assets shall be
valued at the then current depreciated book value
thereof, which value shall be calculated by the
accountancy firm of _______________________
_______________________________________.

The manner of exercise of said option by
CORPORATION shall be as follows:

(1) CORPORATION shall mail or hand deliver to CONTRACTOR its intention to purchase CONTRACTOR'S interest in the concrete plant facility thirty (30) days prior to the proposed purchase date.

(2) CORPORATION shall, immediately upon delivering notice to CONTRACTOR, instruct the accounting firm of ________________________________

_____________________________, to determine the depreciated book value of the concrete plant facility and submit such information to CONTRACTOR and to CORPORATION within thirty (30) days after receiving instructions from CORPORATION. CORPORATION shall, within thirty (30) days after receipt of the calculated depreciated book value, tender to CONTRACTOR the full amount of CONTRACTOR'S interest in the concrete plant facility or shall give to CONTRACTOR a note for the full amount, payable in equal monthly installments over a five-year period, such note to bear interest at the rate of prime plus 1-1/2%.

ARTICLE 15

TERMINATION

This Agreement shall continue until terminated by either party as herein provided. Either party may elect to terminate this Agreement by giving written notice to the other party of its intention to terminate said Agreement. Such notice must be given at least sixty (60) days prior to the proposed termination date. Upon termination of this Agreement, CONTRACTOR shall immediately deliver to CORPORATION all of the records in possession of CONTRACTOR pertaining to the operation of the facility. CONTRACTOR'S continuing right to receive one-third of profits as its management fee shall be terminated concurrent with the termination date of this Agreement, and CONTRACTOR shall also have no further rights to receive the minimum $2,000.00 per month management fee for its services in operating the concrete plant facility.

If CORPORATION elects to terminate this Agreement, CORPORATION may, in its sole discretion, grant to CONTRACTOR the right to purchase CORPORATION'S interest in the concrete plant facility at the then depreciated book value.

Should CONTRACTOR elect to terminate this
Agreement, it must offer to CORPORATION the right
to purchase its interest in the concrete plant facility at
the then depreciated book value.

Any purchase under this ARTICLE 15 shall be
conducted in accordance with the provisions set forth
in ARTICLE 14, subparagraphs (1) and (2).

ARTICLE 16

REMEDIES

Should CONTRACTOR become incapable of
continuing performance of the work herein, whether
due to circumstances within or outside of its control,
CORPORATION may terminate this Agreement.
Should CORPORATION be in default of compensation
owing at any time under this Agreement,
CORPORATION shall be deemed to be in default of
this Agreement, and CONTRACTOR has available to it
all legal remedies and processes.

ARTICLE 17

**ATTORNEYS' FEES
AND COSTS**

If any action at law or in equity is necessary to
enforce or interpret the terms of this Agreement, the
prevailing party shall be entitled to reasonable
attorneys' fees and costs, and necessary disbursements,
in addition to any other relief to which such party may
be entitled.

ARTICLE 18

NOTICES

Any notices to be given hereunder by either party to
the other may be effected either by personal delivery
in writing or by registered or certified mail, postage
prepaid, with return receipt requested. Mailed notices
shall be addressed to the parties at the addresses
appearing at the end of this Agreement, but each party
may change its address by giving written notice in
accordance with this paragraph. Notices personally
delivered shall be deemed communicated as of three
days after mailing.

ARTICLE 19

**ENTIRE
AGREEMENT**

This Agreement supersedes any and all other
Agreements, either oral or in writing, and contains all
the covenants and agreements between the parties.
Each party to this Agreement acknowledges that no
representations, inducements, promises or agreements,
orally or otherwise, have been made by either party or
anyone acting on behalf of any party hereto, which are
not embodied herein, and that no other agreement,

statement or promise not contained in this Agreement
shall be valid or binding. Any modification of this
Agreement will be effective only if it is in writing,
signed by the party to be charged.

ARTICLE 20

**PARTIAL
INVALIDITY**

If any provision in this agreement is held by a Court
of competent jurisdiction to be invalid, void or
unenforceable, the remaining provisions shall
nevertheless continue in full force without being
impaired or invalidated in any way.

ARTICLE 21

**LAW GOVERNING
AGREEMENT**

This Agreement shall be governed by and construed in
accordance with the laws of the State of
_______________________________.

Executed on _________________________________, 19_________, at

_____________________, State of _______________________________.

"CONTRACTOR"

By _______________________________
       President

"CORPORATION"

By _______________________________
       Secretary

<u>**CAUTION**</u>: This form is an illustration only. It has not been prepared with any
specific factual situation in mind. It may not be suitable for your particular use.
No warranty is made as to its fitness for any particular purpose or use. No other
warranty is given. Before using this agreement, or any part of it, you should
consult with your attorney.

This page is intentionally left blank.

AGREEMENT No. 7

COURIER, MESSENGER AND DELIVERY SERVICES

Comment:

This independent contractor agreement deals with another delivery situation. Note the non-assignment clause in paragraph 11. Remember that contracts are generally assignable unless it is stated that they are not assignable. You may not want to contract with A and have him assign the contract to B who you do not want to deal with. However, any restrictions on the worker tend to support employee status. Therefore, you may wish to make your contract freely assignable.

AGREEMENT

AGREEMENT by and between _________________________________ with an

office at ____ hereinafter

called "CONTRACTOR" and ______________   ________________________________,

a _______________________________ Corporation, hereinafter called "CLIENT".

WITNESSETH

WHEREAS, CLIENT is engaged in the air freight forwarding business and for

such purpose requires pickup and delivery service in the city of

____ and its vicinity; and

WHEREAS, the CONTRACTOR, engaged in the business of public transportation,

is willing and able to render such pickup and delivery service as CLIENT requires,

NOW, THEREFORE, it is mutually agreed between the parties as follows:

1. PICK-UP AND DELIVERY OF SHIPMENTS	The CONTRACTOR shall, as an independent contractor, pick up, transport and deliver shipments from the airport(s) serving the aforementioned city to consignees, and from consignors to said airport(s) as requested by CLIENT in connection with CLIENT'S freight forwarding business.
2. LIMITED SERVICE AREA	The geographical scope of the pickup and delivery services provided by the contractor for CLIENT will be limited to (but will not exceed) the air freight terminal area of the airport city shown in the

AGREEMENT as set forth in the tariff identified as
____ or as that
tariff may be revised or reissued from time to time.
No other operating authority is expressed or implied
by the AGREEMENT.

3. MEANS OF CONVEYANCE, LABOR

The CONTRACTOR shall provide and use for such
service means of conveyance which are in the
exclusive charge and control of the CONTRACTOR
and shall furnish an adequate supply of labor which
shall not be subject to the control or supervision of
CLIENT.

4. PAYMENT FOR SERVICE

CLIENT shall pay for the service furnished to it in
accordance with this contract at the following rates:

____

Any adjustment in said rates must be approved by
CLIENT to become effective on the date of the next
scheduled revision to CLIENT'S official air freight
pickup and delivery tariff.

5. COMPLIANCE WITH GOVERN- MENTAL REGU- LATIONS

CONTRACTOR assumes sole responsibility for
compliance with all economic, operational, safety, and
any other requirements imposed by Federal, State,
County, Municipal, or other law or regulator body,
relating to the surface operations hereunder; and
CONTRACTOR agrees to reimburse CLIENT'S entire
costs, including but not limited to the amounts of fines
or penalties and costs of counsel, arising from any
assertion or finding of lack of compliance with the
aforesaid laws and/or regulations in respect to the
surface operations hereunder. CONTRACTOR agrees
that CLIENT may deduct the amounts due it
hereunder from any monies otherwise owed to
CONTRACTOR, in full or partial payment of its
obligation to reimburse to CLIENT its costs as defined
hereinabove.

6. INSURANCE COVERAGE

The CONTRACTOR shall carry and maintain at its
own expense insurance in such forms and amounts as
CLIENT may require and shall furnish to CLIENT a
certificate from all insurance carriers showing the
dates of expiration of any policies, limits of liability
thereunder, and providing that said insurance will not
be canceled or changed until at least thirty days

written notice is given to CLIENT and then only with the consent of CLIENT.

If the CONTRACTOR fails to furnish and maintain such insurance, CLIENT shall have the right to take out and maintain the said insurance for and in the name of the CONTRACTOR and the CONTRACTOR shall pay the cost thereof and shall furnish all necessary information to make effective and maintain such insurance.

MINIMUM LIMITS OF LIABILITY:

1. Workmen's Compensation
  $100,000

2. Comprehensive General Liability Insurance
 Personal Injury
  $100,000 each person
  $300,000 each occurrence
  $300,000 aggregate

 Property Damage
  $100,000 each accident
  $100,000 aggregate

3. Comprehensive Automobile Liability
 Personal Injury
  $100,000 each person
  $300,000 each occurrence
  $300,000 aggregate

 Property Damage
  $100,000 each accident

7. INDEMNIFICA-TION

The CONTRACTOR hereby assumes the entire responsibility and liability in and for any and all damage or injury of any kind or nature whatever to all persons, whether employees or otherwise, and to all property, growing out of or resulting from the execution of work provided for in this contract or occurring in connection therewith, and agrees to indemnify and save harmless CLIENT, its agents, servants and employees from and against any and all loss, expense, including attorneys' fees, damage or injury growing out of or resulting from or occurring in connection with the execution of the work herein

provided for or occurring in connection with or
resulting from the use by the CONTRACTOR, his
agents or employees, or any materials, tools,
implements, appliances, scaffolding, ways or hoists,

elevators, works or machinery or other property of
CLIENT, whether the same arise under the common
law or the so-called worker's compensation law (which
may be in effect at a locality in which work is
situated) or otherwise. In the event of any such loss,
expense, damage or injury, or if any claim or demand
for such damages is made against CLIENT, its agents,
servants or employees, CLIENT may withhold from
contractor any payment due, or hereafter to become
due, under the terms of this contract, an amount
sufficient in its judgment to protect and indemnify it
from any and all such claims, expenses, including
attorneys' fees, loss, damage or injury, or CLIENT in
its discretion, may require the CONTRACTOR to
furnish a surety bond satisfactory to CLIENT
guaranteeing such protection, which bond shall be
furnished by the CONTRACTOR within five (5) days
after written demand has been made therefore.

8. ASSUMPTION OF LIABILITY

CLIENT will assume the liability of the
CONTRACTOR for all risks of loss or damage to
CLIENT'S shipments, including the liability arising
under the CONTRACTOR's agreement to indemnify
CLIENT for any such loss or damage, while such
shipments are in the custody or control of the
CONTRACTOR. In consideration of such assumption
of liability, the CONTRACTOR authorizes CLIENT
to deduct and retain from compensation due the
CONTRACTOR for cartage services rendered, an
amount equal to two percent (2%) for such
compensation.

9. TERMINATION

This agreement may be terminated for any reason by
either party by giving the other party 30 days notice
and may be terminated for cause at any time upon
notice.

**10. CONTRAC-
TOR'S SERVICE
FOR OTHERS**

Nothing herein shall prevent the CONTRACTOR from performing any transportation service for any other person, firm, or company; except that the CONTRACTOR shall not engage in other transportation of air freight without notice to CLIENT.

**11. ASSIGNABIL-
ITY**

No assignment of this contract shall be made by either party without the consent in writing of the other.

**12. CHARGES
COLLECTED BY
CONTRACTOR**

SURETY BOND

In furnishing the services referred to above, the CONTRACTOR may be required to collect upon delivery such charges as directed by CLIENT which will consist of shipper's C.O.D. charges or transportation charges or both. Such charges are to be remitted to CLIENT in accordance with procedures outlined by CLIENT. In those circumstances where funds collected for CLIENT cannot be remitted directly to a CLIENT representative, CONTRACTOR shall obtain, at its expense, a surety bond in the amount of $5,000 and shall furnish to CLIENT evidence that a surety bond is in full force and effect during the term of this agreement.

IN WITNESS WHEREOF, the parties have duly executed this agreement this

_________________ day of _______________________________, 19___________.

CONTRACTOR   CLIENT

By_________________________   By___________________________________

Witness____________________   Witness_______________________________

CAUTION: This form is an illustration only. It has not been prepared with any specific factual situation in mind. It may not be suitable for your particular use. No warranty is made as to its fitness for any particular purpose or use. No other warranty is given. Before using this agreement, or any part of it, you should consult with your attorney.

AGREEMENT No. 8

CARPET INSTALLATION

Comment:

This independent contractor agreement is between a carpet layer or installer and a carpet distributor or dealer. Remember that an agreement will not change an employee into an independent contractor. In some states, such as California, workers engaged in an activity requiring a contractor's license must have their own contractor's license. If not, they are employees of the payer. [See California Labor Code section 2750.5 and California Unemployment Insurance Code section 621.5.]

AGREEMENT

1. IDENTITY OF CLIENT AND IN-DEPENDENT CONTRACTOR

This Agreement is made between _______________ _______________ having a principal place of business at _______________ _______________, hereafter referred to as "CLIENT", and independent contractor having a principal place of business at _______________ _______________, and hereafter referred to as "CONTRACTOR".

2. TERM OF AGREEMENT

This Agreement will become effective on the date stated on this contract for services and will remain in effect until the services provided for herein have been performed.

3. DESCRIPTION OF SERVICES

CONTRACTOR agrees to perform the services described as follows:_______________ _______________ _______________ _______________.

4. MEANS OF PERFORMING SERVICES

CONTRACTOR will determine the methods, details, and means of performing the services described in this Agreement.

5. EMPLOYMENT OF ASSISTANTS

CONTRACTOR may, at CONTRACTOR'S own expense, employ such assistants and helpers as CONTRACTOR deems necessary to perform the services required of CONTRACTOR by this Agreement. CLIENT may not control, direct, or supervise CONTRACTOR'S assistants, helpers, or employees in the performance of those services.

6. PERFORMANCE OF SERVICES

CONTRACTOR shall perform the services required by this Agreement within the period of time defined by the CLIENT.

7. **TERMS OF PAYMENT**

In consideration for the services to be performed by CONTRACTOR, CLIENT agrees to pay CONTRACTOR _________________________________ provided it conforms to CLIENT'S rate of payment for services rendered. Payment is due and payable after completion of the services described in this Agreement to the satisfaction of CLIENT.

8. **TOOLS, INSTRUMEN-TALITIES, AND TRANSPOR-TATION**

CONTRACTOR will supply all tools, instrumentalities, and transportation required to perform the services described by this Agreement.

9. **PROFITS AND LOSSES**

CONTRACTOR is entitled to retain any and all profit he earns from the services he provides pursuant to this Agreement and shall bear at his own expense any and all losses incurred by him in completing the services required of him by this Agreement.

10. **WORKER'S COMPENSATION**

CONTRACTOR agrees to provide workers' compensation insurance to CONTRACTOR'S assistants, agents, helpers, and employees and agrees to hold harmless and indemnify CLIENT for any and all claims arising out of any injury, disability, or death of any of CONTRACTOR'S assistants, helpers, employees, or agents.

11. **INSURANCE COVERAGE**

CONTRACTOR represents to CLIENT that he has a policy of insurance to cover any negligent acts, committed by CONTRACTOR or CONTRACTOR'S assistants, helpers, employees, or agents during the performance of any services required to be performed under the terms of this Agreement.

12. **CONTRACTOR LIABLE FOR COSTS**

CONTRACTOR shall be exclusively liable for all costs incurred by CONTRACTOR in performing the services required of CONTRACTOR by this Agreement.

13. GUARANTEE OF WORK AND SERVICES

CONTRACTOR guarantees for a period of one year from the date of the completion of the services required of CONTRACTOR by this Agreement that said work and services shall be free from any and all defects. CONTRACTOR agrees that as to any and all defects occurring within one year of the date of the completion of the services required of CONTRACTOR by the Agreement, that CONTRACTOR shall make all repairs necessary to correct said defects and that CONTRACTOR shall be solely liable for all costs and expenses incurred by him in repairing said defect or defects. CONTRACTOR agrees that if he fails to make said repairs to the satisfaction of CLIENT within seven (7) days after being notified, either orally or in writing, of the defect or defects that CLIENT is entitled to contract with another party to make the necessary repairs, and CONTRACTOR agrees to reimburse CLIENT for all costs and expenses incurred by CLIENT in contracting with another party for purposes of making the aforementioned repair or repairs.

14. ENTIRE AGREEMENT

CONTRACTOR acknowledges that CLIENT has made no representations to CONTRACTOR except as set forth in this Agreement. This is the entire agreement between the parties.

IN WITNESS WHEREOF, the parties have duly executed this agreement this

__________________________ day of __________________________ , 19________ .

CONTRACTOR   CLIENT

By:_________________________   By:_________________________________

CAUTION: This form is an illustration only. It has not been prepared with any specific factual situation in mind. It may not be suitable for your particular use. No warranty is made as to its fitness for any particular purpose or use. No other warranty is given. Before using this agreement, or any part of it, you should consult with your attorney.

AGREEMENT No. 9

BARBERY AND COSMETOLOGY LEASE

Comment:

This Lease Agreement deals with the barbery and cosmetology business. With respect to employment tax audits it is, as a general rule, safer to be a payee then a payer. In a lease agreement, the payee or shop owner is the landlord. A landlord-tenant relationship is to a great extent inconsistent with an employer-employee relationship.

AGREEMENT

This Lease made _____________________, 19___, between _____________________

_______________________________________individually and doing business at

_____________________________, City of _____________________, County of

_________________, State of _________________, herein referred to as "LESSOR,"

and _____________________________ of _____________________ City of

_______________________________________County of _____________________, State

of _____________________________, herein referred to as "LESSEE."

WHEREAS, LESSOR is the sole owner of the business premises described below

having space therein to let; and

WHEREAS, LESSEE is in the business of barbery and cosmetology and desires to

lease space from LESSOR; and

WHEREAS, the parties desire to enter a Lease Agreement defining their respective

rights, duties and liabilities relating to the premises.

In consideration of the mutual covenants contained herein, the parties agree as

follows:

SECTION 1.   The Lease shall be on a month-to-month basis. LESSEE
TERM OF LEASE   shall surrender the premises to LESSOR immediately
upon termination of the Lease.

<table>
<tr><td valign="top">

SECTION 2.
RENTAL

</td><td valign="top">

LESSEE agrees for himself, his heirs, executors, administrators and assigns to pay in the manner hereinafter specified to LESSOR and his heirs and assigns, without demand, rent for the demised space. During the term hereof, LESSEE shall pay to LESSOR a minimum daily rental, payable daily, a sum equal to thirty percent (30%) of the daily gross receipts of the business done by LESSEE; provided further however, the minimum rental per month shall be Three Hundred Dollars ($300.00) payable on the last day of each and every month during the term hereof.

Payments shall be deemed and considered as rental of the demised premises and it is expressly understood by the parties hereto that this provision is strict rental only, and shall in no way be considered or construed as creating the legal relation of a partnership, and it is further expressly understood that LESSOR is in no way responsible for any losses which LESSEE may sustain at any time.

If LESSOR is compelled to incur any expenses, including reasonable attorneys' fees, in instituting and prosecuting any action or proceeding by reason of any default of LESSEE hereunder, the sum or sums so paid by LESSOR, with all interest, costs, and damages shall be deemed to be additional rent hereunder and shall be due from LESSEE to LESSOR on the first day of the month following the incurring of such respective expenses.

</td></tr>
<tr><td valign="top">

SECTION 3.
RESTRICTIONS ON
USE

</td><td valign="top">

LESSEE shall not use or permit the premises, or any part thereof, to be used for any purpose other than those set forth herein. LESSEE shall neither permit on the premises any act, sale, or storage that may be prohibited under standard forms of ____________ _________________ insurance policies, nor use the premises for any such purpose. In addition, no use shall be made or permitted to be made that shall result in, (1) waste on the premises, and, in fact, shall be obligated to perform such janitorial and cleaning services as shall be required to eliminate such waste caused directly by LESSEE, (2) a public or private nuisance that may disturb the quiet enjoyment of other tenants in the building, (3) improper, unlawful or objectionable use, including sale, storage, or preparation of food, alcoholic beverages, or materials generating an odor on the premises, or (4) noises or

</td></tr>
</table>

vibrations that may disturb other tenants. LESSEE
shall comply with all Government regulations and
statutes affecting the premises, either now or in the
future. LESSEE shall also comply with the rules and
regulations to which LESSOR is subjected under his
lease with the prime LESSOR.

SECTION 4.
LICENSES

LESSEE is solely responsible for applying for,
obtaining, and maintaining all licenses, permits,
applications, and certificates required by law in the
operation of LESSEE'S business on and in LESSEE'S
leasehold.

SECTION 5.
UTILITIES

LESSOR shall furnish all heat and air conditioning to
the demised premises on all business days during the
appropriate season.

LESSOR shall furnish all electricity required by
LESSEE in the normal conduct of business activities on
the premises, but LESSOR shall be entitled to review
the proposals of LESSEE to add any equipment
requiring large electrical power supplies, and to charge
LESSEE for the additional costs of the increased
electrical service, if LESSOR deems the necessity
therefore reasonable.

LESSOR shall furnish all hot and cold water required
in the normal conduct of business.

SECTION 6.
CREDIT CARDS,
CHECKS,
CASHIER, REFER-
RALS

LESSOR shall furnish one credit card recording
machine for use by LESSEE. For the sole purpose of
LESSEE'S convenience, LESSEE may pay his weekly
rental in credit card receipts and/or checks paid to
LESSEE.

LESSOR shall furnish a cash register for the use and
convenience of LESSEE, but the LESSEE is responsible
for the maintenance, control, and disbursements.

**SECTION 7.
ALTERATIONS
AND MODIFICA-
TIONS; REPAIRS**

LESSEE has inspected the premises, and they are now in a tenantable and good condition. LESSEE shall take good care of the premises and shall not alter, repair or change the premises without the written consent of LESSOR. All alterations, improvements, and changes that LESSEE may desire shall be done either by or under the direction of LESSOR, but at the expense of LESSEE and shall become the property of LESSOR and remain on the premises, except that at the option of LESSOR, LESSEE shall, at its expense, remove from the premises all partitions, counters, railings, and similarly installed improvements when surrendering the premises. All damage or injury done to the premises by LESSEE or any person who may be in or on the premises with the consent of LESSEE shall be paid for by LESSEE. LESSEE shall, at the termination of this Lease, surrender the premises to LESSOR in as good condition and repair as reasonable and proper use thereof will permit.

LESSOR shall be responsible for making all routine repairs and for performing routine maintenance.

**SECTION 8.
DAMAGES; MAL-
PRACTICE INSUR-
ANCE; WORKMEN'S
COMPENSATION
INSURANCE**

LESSEE agrees to pay for all damage to building, as well as all damage or injury suffered by tenants or occupants thereof caused by misuse or neglect of the premises by LESSEE.

LESSEE agrees to maintain malpractice insurance on himself, his agents and employees in an amount in excess of twenty thousand ($20,000) dollars. LESSEE further agrees to maintain workmen's compensation insurance on his agents and employees, if any.

**SECTION 9.
DESTRUCTION OF
PREMISES**

In the event of a partial destruction of the premises during the term of this Lease, from any cause, LESSOR shall forthwith repair the same, provided the repairs can be made within thirty (30) days. Any partial destruction shall neither annul nor void this Lease.

**SECTION 10.
ARBITRATIONS**

Any dispute between LESSOR and LESSEE relative to the provisions of this Lease shall be subject to arbitration. Each party shall select an arbitrator, and the two arbitrators so select a third arbitrator between them, the controversy being heard by the three arbitrators so selected. The decision of the three arbitrators shall be final and binding on both LESSOR and LESSEE, who shall bear the cost of such arbitration equally between them.

<table>
<tr><td>

**SECTION 11.
CONDEMNATION**

</td><td>

A condemnation of the entire building or a condemnation of the portion of the premises occupied by LESSEE shall result in a termination of this Lease agreement. LESSOR shall receive the total of any consequential damages awarded as a result of condemnation proceedings. All future rent installments to be paid by LESSEE under this Lease shall be terminated.

</td></tr>
<tr><td>

**SECTION 12.
ASSIGNMENT AND
SUBLEASE**

</td><td>

LESSEE shall not assign any rights or duties under this Lease nor sublet the premises or any part thereof, nor allow any other person to occupy or use the premises without the prior written consent of LESSOR.

</td></tr>
<tr><td>

**SECTION 13.
BREACH OR
DEFAULT**

</td><td>

LESSEE shall have breached this Lease and shall be considered in default hereunder if, (1) involuntary proceedings are instituted against LESSEE under any Bankruptcy Act, (2) LESSEE fails to pay any rent when due, or (3) LESSEE fails to perform or comply with any of the covenants or conditions of this Lease and such failure continues for a period of three days after receipt of notice thereof from LESSOR.

</td></tr>
<tr><td>

**SECTION 14.
EFFECT OF
BREACH**

</td><td>

In the event of a breach of his lease as set forth in Section Thirteen, the rights of LESSOR shall be as follows:

(1) LESSOR shall have the right to cancel and terminate this Lease, as well as all of the right, title, and interest of LESSEE hereunder, by giving to LESSEE not less than three (3) days notice of the cancellation and termination. On expiration of the time fixed in the notice, this Lease and the right, title, and interest of LESSEE hereunder shall terminate in the same manner and with the same force and effect, except as to LESSEE'S liability, as if the date fixed in the notice of cancellation and termination were the end of the term herein originally determined.

</td></tr>
</table>

(2) LESSOR may re-enter the space immediately and remove the property and personnel of LESSEE, and store the property in a public warehouse or at a place selected by LESSOR, at the expense of LESSEE. After re-entry, LESSOR may terminate the Lease on giving three (3) days written notice of termination to LESSEE.

After re-entry, LESSOR may relet the space or any part thereof for any term without terminating the Lease, at the rent and on the terms as LESSOR may choose.

SECTION 15. UNLAWFUL DE-TAINED AND AT-TORNEYS' FEES

In case suit shall be brought for an unlawful detained of the premises, for the recovery of any rent due under the provisions of this Lease, or for the LESSEE'S breach of any other condition contained herein, LESSEE shall pay to LESSOR a reasonable attorneys' fee which shall be fixed by the court, and such attorneys' fee shall be deemed to have accrued on the commencement of the action and shall be paid on the successful completion of this action by LESSOR.

LESSEE shall be entitled to attorneys' fees in the same manner if judgment is rendered for LESSEE.

SECTION 16. TAXES

LESSEE agrees for himself, his heirs, executors, administrators and assigns to pay all taxes, assessments, penalties, etc., arising out of the operation of LESSEE'S business; included, but not limited to, federal and state income tax, sales tax, social security taxes, unemployment taxes, etc. Such obligation shall include all prepayments and advance payments required by law. Such payments shall be made sixty (60) days before the final due date and a copy of evidence of payment shall be forwarded to LESSOR.

LESSOR is in no way or manner obligated to pay any of the aforesaid charges on LESSEE'S behalf.

SECTION 17. REMEDIES OF LESSOR CUMULA-TIVE

The remedies herein given to LESSOR shall be cumulative, and the exercise of any remedy by LESSOR shall not be to the exclusion of any other remedy.

IN WITNESS WHEREOF, the parties have executed this Lease at

____

___________________, State of _________________________, the day

and year first above written.

LESSOR   LESSEE

By:_________________________   By:___________________________

STATE OF _______________)
)
COUNTY OF ____________)

On ____ before me, the

undersigned, a Notary Public in and for said State, personally appeared

____,

personally known to me or proved to me on the basis of satisfactory evidence to be

the person(s) whose name _________________________________ subscribed to the

within instrument and acknowledged that _____________________________

executed the same.

        WITNESS my hand and official seal.

____
        Notary Public in and for said County

                and State

This page is intentionally left blank.

AGREEMENT No. 10

PLUMBING SERVICES

Comment:

This independent contractor letter agreement deals with the relationship between an independent plumber and a plumbing company. Caveat: In many states (such as California) plumbers need contractor's licenses. Without such a license, if required, it may be hard to contend that the plumber is an independent contractor. This letter agreement contains language attempting to affect unemployment compensation rights. If the worker is in fact an employee, the language will have no impact on any unemployment compensation rights. This letter agreement is from the independent contractor. It can, of course, be changed to make it come from the client.

INDEPENDENT PLUMBER
2000 - 5th Avenue
Northtown, NY 10595

January 3, 1989

Mr. John Smith, President
ACME PLUMBING COMPANY
100 Main Street
Anytown, NY 10595

     Re: Independent Contractor Agreement between _______________________

____ and

____

Dear Mr. Smith:

I wish to enter into an independent contractor agreement with your company.

Pursuant to such an agreement it is understood that I will be treated as an independent contractor and not as an employee of your company. I will personally, and on my own accord, be responsible for the payment of my federal income taxes, both estimated income tax and self-employment tax; and my state income taxes.

At the conclusion of this agreement it is understood that I will not file for unemployment benefits with either the federal or state government as I am not entitled to any. It is understood that should a dispute arise as to any payment due

74

under this agreement that I will not seek to enforce any payment hereunder

through the Division of Labor Standards Enforcement of the State of

_______________________ but will seek any civil means available to me.

You have agreed that you will not exercise control over the means of my

work and that your only concern is that the result of my work be completed to the

standard set forth in the trade of plumbing. Further, I agree that my work shall

be to the satisfaction of not only your company but to the land owner or general

contractor of the project.

It is understood that should any defect in workmanship arise that I will

be responsible for the repair of said defect and if not, that my payment hereunder

will be withheld until the completion of repair of said defect. Should your

company have to seek other means of repair, the cost of such a repair will

automatically be deducted from any sum due hereunder.

It is further agreed that I will furnish the necessary tools and equipment

to complete my work.

In regard to workmen's compensation insurance, it is understood that

your company will cover me for said insurance but that this coverage shall in no

way indicate a change in my status as an independent contractor.

It is further understood that you will file all required government forms

that will indicate the amount of my payment hereunder as required by law.

[Continued on Next Page]

Upon acceptance of these terms by your company, this letter shall constitute an agreement between the parties pursuant to the terms of this letter.

Sincerely,

INDEPENDENT PLUMBER

By: _______________________________
Harold Jones, Sole Proprietor

AGREED TO THIS _________________ **DAY OF** _____________________, 19____

ACME PLUMBING COMPANY

By: _______________________________
John Smith, President

<u>CAUTION</u>: This form is an illustration only. It has not been prepared with any specific factual situation in mind. It may not be suitable for your particular use. No warranty is made as to its fitness for any particular purpose or use. No other warranty is given. Before using this agreement, or any part of it, you should consult with your attorney.